Arado
Ar 196

Germany's Multi-Purpose Seaplane

Ar 196 A-2, CU+AD is transported to its trial flight.

Hans-Peter Dabrwoski/Volker Koos

SCHIFFER MILITARY HISTORY
Atglen, PA

NP+ZO during factory testing in Warnemünde.

Photos:
P. Petrick
V. Koos
H.J. Meier
H.J. Nowarra
Hefner, Wenzel, Drüppel via Petrick
B. Lange

Translated from the German by James C.Cable.

Copyright © 1993 by Schiffer Publishing Ltd.

Printed in the United States of America.
ISBN: 0-88740-481-2

This title was originally published under the title,
See-Mehrweckflugzeug Arado Ar 196,
by Podzun-Pallas Verlag, Friedberg.

We are interested in hearing from authors with book ideas on related topics. We are also looking for good photographs in the military history area. We will copy your photos and credit you should your materials be used in a future Schiffer project.

Published by Schiffer Publishing, Ltd.
77 Lower Valley Road
Atglen, PA 19310
Please write for a free catalog.
This book may be purchased from the publisher.
Please include $2.95 postage.
Try your bookstore first.

We are interested in hearing from authors
with book ideas on related subjects.

PREFACE

The origin of Arado aircraft can be traced back to the establishment of the Friedrichshafen Flugzeugbau GmbH on June 12, 1912. Even during the First World War, this factory delivered about 40% of all German seaplanes. The plant on the Bodensee lake was later complimented with a branch facility in Warnemünde near Rostock. This is where the floats and wings as well as about 50 complete FF 49c aircraft were constructed beginning in early 1918.

After World War I, Arado built smaller aircraft, and starting November 1920 the Warnemünde plant — now under the leadership of the Stinnes concern — was renamed as a branch subsidiary of the "Dinos"-Automobilwerke AG. Beginning in June 1925, after a re-designation to Arado Handelsgesellschaft mbH Werft Warnemünde, they started to build aircraft once more, even if only under license of the Heinkel firm to build the HD 21 (15 examples) and the HD 32 (4 examples). Beginning in 1926, Arado began manufacturing its own aircraft, and had built about 35 by 1933, mostly training aircraft as well as single-model aircraft for various special purposes. The approximate 100 employees also continued to build small aircraft, furnishings as well as bombing mechanisms, all for the secret arming of the Reichswehr.

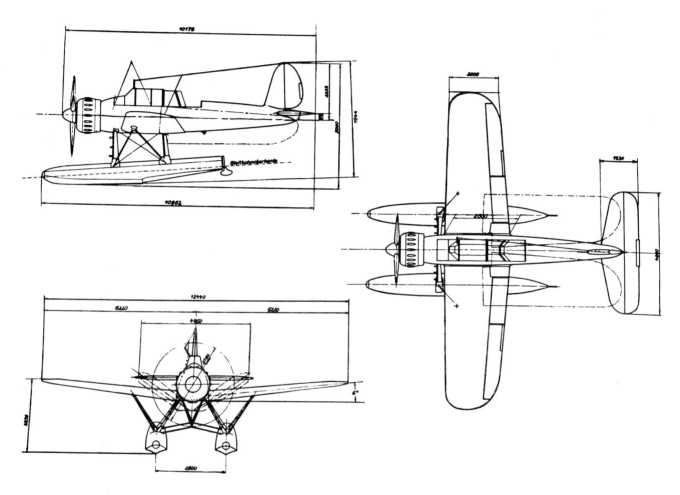

Above: Three-sided view of the Arado Ar 196 A.

Right: Ar 196 V1 in its first version. The tail number license is D-IEHK. Note the horn balance on the rudder.

Above: V1 in a later version, with the tail number **D-IEHK** remaining, but the horn balance is now missing from the rudder.
Below: V1 taking off on the Warnow.

After the not-entirely-voluntary merger of the Focke-Wulf and Albatross concerns in 1930, Chief Builder Dipl.Ing. Walter Blume, together with seven of his most capable co-workers, came to Arado and became the director of the construction bureau and later the technical director.

From March 4 1933 on, the firm was called "Arado Flugzeugwerke GmbH" and from that point on built only aircraft. The production facilities in Warnemünde were greatly enlarged, and even got its own factory airfield in 1934. The number of employees grew to 6,000 by 1943, and the main factory was in Brandenburg beginning in 1935, thereby causing the Warnemünde facility to lose its importance. Branch factories were established in Anklam, Babelsberg, Eger, Rathenow and Wittenberg, among others. Mass-production of Heinkel and Messerschmitt models under license made up the majority of the output. Noteworthy models of their own production which should be mentioned are the Ar 66 training aircraft and the Ar 196 multi-purpose seaplane. Most of the other Arado designs were only built singly or in small numbers and, with a few exceptions (such as the Ar 96 training aircraft and the Ar 234 turbo-jet bomber/reconnaissance aircraft) remained widely unknown.

DEVELOPMENT & CONSTRUCTION OF THE Ar 196

Above and below: The V1 received a three-bladed propeller and a flat canopy for its planned record attempt.

After the He 114, which was the successor model to the Heinkel He 160, had proven itself insufficient in the early stages of testing as a shipborne aircraft

Above: Ar 196 V2, D-IHQI, serial number 2590.
Below: The same photograph, given a military retouching by the propaganda department.

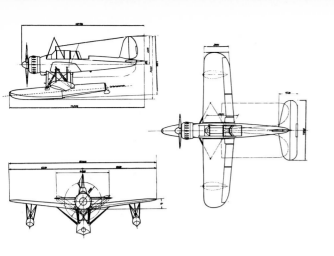

The B version of the Ar 196 had a centrally-mounted float under the fuselage and a small support float under each wing. V3 D-ILRE, V4 D-OVMB, V5 D-IPDB.

for the *Kriegsmarine*, the technical bureau of the RLM issued a new invitation to bid for a new two-seat reconnaissance aircraft capable of catapult launch. Beginning in autumn 1936, the firms of Focke-Wulf and Arado built suitable prototype models in response. These were, in accordance with the demands of the RLM, each outfitted in two variants — as two-float and as center-float aircraft, for which Focke-Wulf proposed a conventional biplane, while Arado envisioned a low-wing monoplane.

The two firms built four prototypes each. Among these were the first two double-float aircraft (Ar 196 V 1, serial number 2589, D-IEHK / Ar 196 V 2, ser. no. 2590, D-IHQI / Fw 62 V 1, ser. no. 2062, D-OFWF / Fw 62 V 2, ser. no. 2063, D-OKDU) and those that followed received the central float (Ar 196 V 3, ser. no. 2591, D-ILRE / Ar 196 V 4, ser. no. 2592, D-OVMB / Fw 62 V 3, ser. no. 2064, D-OHGF/ Fw 62 V 4, 2065, D-OMCR).

The Ar 196 V 1 flew for the first time in summer 1937, while Focke-Wulf was only able to do this on October 23rd of that year.

Comparison testing of both models took place in the first quarter of 1936 at the Travemünde testing facility. The first order for 10 pre-production aircraft of the two float Ar 196 A-0 model equipped with the BMWz132 K engine was delivered beginning in November 1938. Although the Fw 62 proved itself superior to the Arado model during water testing and in the area flight characteristics, the latter was ordered in February 1938, even before the end of the testing. Conclusive features in favor of this decision were the more modern and simpler construction techniques and the aircraft's higher load capacity.

There had still not been a definitive assessment of the advantages or disadvantages of the two different float systems. Arado still delivered a fifth prototype (Ar 196 V 5, serial number 0090 (D-IPDB) with a centrally mounted float and side-mounted support floats, the so-called "B" version. Further testing was done at the facility at Travemünde, during which the float suspension utilized on the Fw 62 was also closely examined. These measures taken to improve the handling characteristics on the water were already in use on several foreign models, but had been tested by Focke-Wulf on the W 7 bi-winged seaplane developed for the Reichswehr in 1930/31, and by the Gothaer Waggonfabrik on several float planes in World War I.

In the end, the double-float model Arado 196 was the version to go into mass production.

Above: Assembly of the Ar 196.

Below: A completed A2 with a pre-war light grey color scheme.

The exact purpose of the Ar 196 V 6 (serial number 0191), which came into service on 23 September 1941, cannot be determined even to this point. This aircraft entered service three years later with the *Ergänzungs-Bordfliegerstaffel*. After testing by the military of the pre-production model as a ship-launched reconnaissance aircraft with its maneuverable MG 15 machine gun for defense and two 50-kilogram bombs under the outer wings for attack purposes, Arado delivered the first 20 mass-produced aircraft Ar 196 A-1 beginning in June 1939. The first units to receive the aircraft were the 1. and 5. *Staffel* of *Bordfliegergruppe 196*. By the end of 1939, the battleships of the *Kriegsmarine* had up to four of these aircraft on board, which were also supplied to *Kreuzer* (cruisers) and *Hilfskreuzer* (auxiliary cruisers) warships.

The next series produced was the Ar 196 A-2, which had a supplemental fixed attack weapons, consisting of two 20mm MG-FF/M cannons in the wings and a 7.9mm MG 17 machine gun in the right side of the fuselage. This more potent version came into use as a coastal support reconnaissance aircraft and was delivered by the Warnemunde Arado plant beginning November 1939. The attack weapons were tested for the first time in the fourth prototype model.

The number of those Ar 196 A-2's constructed remained small, as was the case with the Ar 196 A-4, which was delivered starting at the end of 1940 as a replacement for the Ar 196 A-1 of the *Bordaufklärerstaffeln* and had the improved weaponry of the A-2 and better radio equipment as well.

The chief production model was the Ar 196 A-3 built beginning in 1941 at the Warnemünde

Above: Ar 196 A-0 D-IYFS. Below: A-01 D-ISFD (later T3+HB) of the 10./(See) LG2 during testing on the Atlantic near Travermünde in 1939.

An Ar 196 of *Lehrgeschwader 2.*

facility and at the French S.N.C.A. plant in St. Nazaire from July 1942 to March 1943, which delivered 23 aircraft.

The last production model was the Ar 196 A-5, produced beginning in 1943 in Warnemünde, where production came to an end in March 1944. These variants were also built in the Fokker plants in Amsterdam, which produced a total of 96 aircraft in the course of one year up to August 1944. That was the end of the production of the most successful German seaplane of World War II, after a total of 600 such aircraft had been produced.

CONSTRUCTION & SERIES OVERVIEW

The Arado 196 is a single-engine seaplane which can be catapult launched, and is equipped with a BMW 132 K engine. Its mission is reconnaissance and engagement of small ship units.

The fuselage, consisting of a welded steel tube frame with trapezoid shaped cross-section, which was made round by bent tubing, longitudinal battens and metal spanners, was covered in the forward portion with aluminum, the remainder covered with fabric.

The 9-cylinder air-cooled BMW 132 K rotary engine is fastened to the fuselage frame with four spherical mounts. It had a low blower. The engine's sheet metal covering had aerodynamic bulges for the valve arms, a trimming and access hatches for maintenance.

Under the pilot's seat, which could be adjusted for height, was the nautical equipment bay with lines and a sea anchor. The observer's seat was fastened to rollers and could be adjusted along rails and clicked into three different positions. Both seats were protected by a jettisonable plexiglass canopy, which was open in the rear.

The trapezoid-shaped all-metal wings have two spars and 32 ribs. Quickly disconnecting wing mounts permitted the wings to be quickly folded to the rear along the fuselage, and then utilize a V-shaped auxiliary brace between the upper side of the wing and the upper chord of the fuselage for support. The horizontal and vertical stabilizers were of all-metal construction, the respective control surfaces were of light metal construction covered with fabric. The control surface were operated by rods, levers and control lines.

The floats of the mass-produced version of the Ar 196 were two interchangeable all-metal floats made of aluminum, each with a 2,750-liter volume. The had a keel and a step. There were eight tie-

Preparations for catapult launch. Above: The wings of this aircraft are still folded. The support braces can be clearly seen.
Below: Ready for take-off.

Above: An Ar 196 serving on board the battleship *Gneisenau*.
Below: Ready for take-off on the catapult.

downs on the rounded deck. In each of the floats there was a sack for extra fuel, which had a 300 liter capacity, a smoke generator and an area containing emergency ammunition and rations which could be opened from above and below.

The larger center float of the B-version had principally the same construction, but had a 4,535-liter volume and had 13 tie-downs. Built into this float there was a 600-liter fuel bag, two smoke generators and one emergency food and ammunition container. The two side-mounted support floats were also constructed of hydron aluminum. The main floats of both basic versions have a rudder on the rear which could be operated from the cockpit.

The following production series are known: according to a production series overview from 1 November 1942:

Ar 196 A-2: engine: BMW 132 K, armament: 1 maneuverable MG 15 machine gun, with 525 rounds of ammunition, radio: 1 FuG V AU for bombing: 2 ETC 50 VIII See, propeller: 3-blade

Right: The antenna mount on the underside of the fuselage.
Below: Servicing the aircraft — essential for incident-free operation.

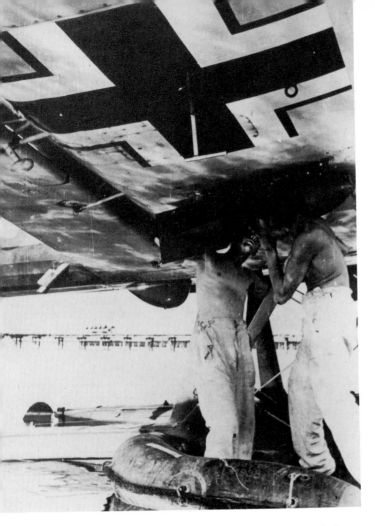

When used in armed operations, the Ar 196 carried two 50-kilogram bombs which are being loaded here and installed into the *ETC*'s.

variable-speed Schwarz

Ar 196 A-2: like the A-1, but with two **FF/M** machine guns and one **MG 17** and partly variable propeller

Ar 196 A-3: like the A-2, but with **FuG X** and **Peil G IV** radio equipment and variable-speed propeller

Ar 196 A-4: like the A-2, but with 800 liters of fuel.

Ar 196 A-5: like the A-3 but with radios **FuG X**, **FuG 25** and the **FuG 6** set, the **ASK-RZ** bombing system for ripple-bombing, **MG 81 Z** machine gun for self-defense

The production overview dated 1 November 1942 credits the Ar 196 A-2 with an additional **SN 5 C** smoke generator and the A-3 with a VDM variable-speed propeller, artificial horizon and retro-fitting of the MGz81 Z machine gun in the observers position.

Two years later, on 1 November 1944, the following additions were made:

Ar 196 A-3/U 1: like the A-3 but with snow skis

Ar 196 A-4/U 1: like the A-2 but with towing apparatus

Ar 196 A-5: BMW 132 W engine, 2 MG 131 or MG 151 machine guns instead of the earlier wing-mounted guns (MG/FF).

USE & DEPLOYMENT OF THE Ar 196

At the beginning of the war on 1 September 1939,

After the flight testing came the operational troop units.
Above: The rear portion of the canopy.
Below: Weapons maintenance (1./Kü.Fl.Gr. 706, Aalborg 1941).
Above right: The writing above the wing root states, "Air/sea rescue equipment container."
Lower right: Clearly visible is the positioning of the weapons which fire through the propeller arc.

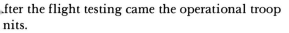

Left: Preparing for take-off. Here, an Ar 196, T3+LK of 2./Bordflg.Gr. 196, in May 1941.

Right: The aircraft is lowered to the water.

Left: The aircraft takes off from calm waters on a reconnaissance flight.

Above: In the foreground is serial number 1960056 with markings on the vertical stabilizer denoting ships sunk, and in the background is 6W+AN of 5./Bordflg.Gr. 196.
Below: Headquarters aircraft of II./Kü.Flg.Gr. 706.

Above: Larger ship units took on up to three Ar 196's. Here is an assembly of those belonging to Bordflg.Gr. 196.

Below: T3+IH of 1./Bordflg.Gr. 196

Maintenance and flight preparations for seaplanes are somewhat more involved than those on land — as one can see here.

Above: Aalborg, 1941. Behind the two Ar 196's in the foreground is an He 114. Below: Testing the engine of T3+FH belonging to 1./Bordflg.Gr. 196.

Right: 7R+BK of 2./SAGr.125, showing clear signs of use, is freed from its camouflage netting.

Left: Ar 196's of the 2./Bordflg.Gr. 196 in Kiel-Holtenau.

Right: 6W+EN of the 5./Bordflg.Gr. 126 is made fast.

Above: An Ar 196 flying escort for a Junkers Ju 52 over the Mediterranean.
Below: An aircraft belonging to 2./SAGr. 125 in the Mediterranean theater of operations.

Right: CK+FC —
The pilot and radio-
operator are carried
to the aircraft on
piggy-back.

Left: After returning
from an operational
sortie, the air crew
are carried back to
land in the same
manner (Atlantic
coast).

Right: DN+IF, serial
number 0226, is
made fast once the
floats are resting on
firm ground.

The reconnaissance aircraft departs the battleship via catapult. After the flight, it lands on the ship's so-called wake and is then hoisted back into place by a crane on deck.

Here we see an Ar 196 hanging from a crane (6H+LM of Flg.Erg.Gr. (See), Kamp).

Above and below: **CK+EQ** in Norway.

DH+HT on the ramp of the Warnemünde Arado factory airfield. The braces used when folding the wings have not yet been removed.

A propaganda photo montage: An Ar 196 forces the British submarine "Seal" to surface.

True? "Two Arado's victorious in combat over a Whitley" (*Der Adler*, 2 March 1943).

the 16 *Seefliegerstaffeln* and three *Trägerstaffeln* of the *Kriegsmarine* had a total of 154 seaplanes, of which eight were Ar 196's, which served in the 1st *Bordfliegerstaffel 196* in Wilhelmshaven. These aircraft were assigned to the largest ship units, where they relieved the old He 60's. At the end of the year, the 5./B.Fl.St. 196 in Holtenau was also equipped with Ar 196's. With the widening of the combat theater to danish and norwegian waters and the Belgian-French Atlantic coast in 1940, both *Bordfliegergruppen* were moved several times, split up and assigned to various command locales. Their main mission was the reconnoitering and engagement of submarines.

One engagement which was particularly cannibalized by propaganda was the locating and forcing to surface of the British submarine "Seal" on May 5, 1940 by Arado 196's of the 5./195 stationed in Aalborg.

During the course of 1940 further naval aviation in addition to the *Bordfliegerstaffeln* received the Arado low-winger. Among these were the *Bordflieger-Ergänzungsstaffel*, which operated under the name of 3./Fl.Erg.Gr. (See), and the first *Staffel* of *Küstenfliegergruppe 706*. Of those 251 seaplanes lost from the end of 1940 to war's end, 32 were Arado 196's.

Again and again numerous other seaplane units received aircraft of the various Arado types, which performed their duties on all coastlines and seas occupied by German troops from the North Atlantic to the Black Sea. As such, the *Seeaufklärungsgruppen* (maritime reconnaissance) 125, 126, 127, 128, 130, 131 and 132 all had at least one or more *Staffeln* or *Ketten* equipped with Arado 196's. The Arado 196 was the most-built German seaplane and estimated to be the Luftwaffe's best float plane. At the end of March 1944, at the time of the largest coastal and pre-coastal extension of German power, seven *Küstenfliegerstaffel* flew the Ar 196. Of these 64 aircraft, the majority, specifically 46 aircraft, flew in the Aegean and Black Seas. But the Ar 196 was also present in the Far East. In addition to those which were there on board the *Hilfskreuzer* ships, the *Fliegerkommando* of the navy's special duty station at Penang had two Ar 196's for protection of the German U-boats stationed there.

The mission of the Ar 196 was multi-faceted. In addition to reconnaissance of the coastline and protecting larger ship units, attacking smaller ship units, enemy anti-submarine reconnaissance aircraft, courier duty, sea rescue duty and engaging partisan units in the Aegean were all part of the daily duties of this robust aircraft.

Right: Small
repairs on
DN+IG.

Left: DN+IE,
serial number
1960275.

Right: DN+IF,
serial number
1960276.

Above and below: Scenes from flight trials. The aircraft are towed to the take-off point . . .

. . . and then lift off, as **DN+IG** does here, on its test flight.

Aircraft DH+HT is taken out of Hall H of the Arado Flugzeugwerke plant and is towed by a tractor across the plant grounds to the ramp area.

Manufacture of an Arado in Warnemünde.

Left: An Ar 196 is made ready for take-off aboard the battleship *Prinz Eugen* (T3+BL).

Right: Catapult launch from the *Prinz Eugen*.

Left: T3+AM is taken aboard once again.

Above: T3+GH seen with wing braces, below: an aircraft of 3./Bordflg.Gr. 126 seen in the background.

GA+DX, serial number 0124, was used to make many factory and propaganda. Below: Ar 196's of SAGr.126 on the Danube river near Vokovar (in the foreground is D1+BN).

Ar 196 **BB+YQ**, serial number 0143, brand-spanking new and ready for operations.

Above: DN+IE, serial number 0225, is pushed onto calm water. Below: DH+?? stands ready on the bank.

OU+AR returns from a flight, below: an Ar 196 practices approaching targets.

Deployment from auxiliary cruisers is difficult. Often the Ar 196 had absolutely no markings, as with this aircraft serving aboard the auxiliary cruiser *Widder*.
Below: An aircraft of the cruiser *Orion* flipped on take-off on October 2, 1940 and was recovered, though severely damaged.

Here is more damage: The aircraft above was stricken by a south-sea swell which tore out the engine and its mounts.

Below: the newspaper *Wehrmacht* reported on 28 July 1943 that this aircraft tipped over while landing on rough seas.

The end: PO+HH in Keil-Holtenau, May 1945.

Above: A Bulgarian Ar 196 on the Chaika (near the Black Sea).
Below: One of the Bulgarian aircraft later in the Varna museum.

Above: AirMin 92 — one of the two aircraft captured by the British.

Below: An Ar 196 captured by the Americans seen here with the "Iron Cross" (!) at Willow Grove Naval Air Station, Pennsylvania.

The Ar 199, which look fairly similar to the Ar 196, should also be mentioned here in brief. About fifteen of these training aircraft were produced.
Above: V1, serial number 3671, NH+AM, formerly D-IRFB.
Right: A three-sided view.

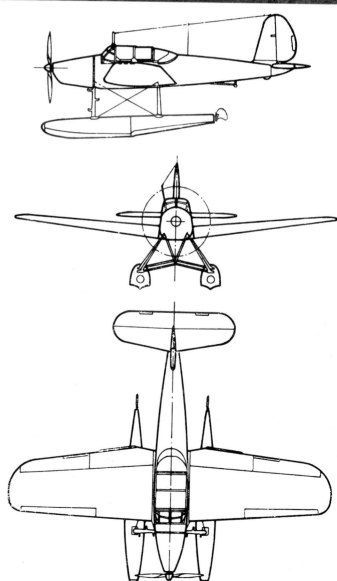

As a ship-launched airplane on board *Hilfs-kreuzern*, which were conducting anti-maritime shipping operations in the Pacific and Indian Oceans the Ar 196 flew their missions without any particular markings other than a light blue paint job. These "free pirates of the sea" used camouflage as a method of attack. In order to prevent the ship under attack from reporting its position, the Ar 196 utilized a special "Antenna Cutter" which consisted of a steel cable with a weighted end, somewhat like a free-hanging long wire antenna. When wound up, it hung in a cylinder-shaped container in the rear of the fuselage. After a ship had been detected, the aircraft assumed a low altitude, unrolled the cable which had three sharpened edges and towed at an acute angle, and flew at the ship from broadsides. As the aircraft overflew the enemy the antenna cutter was pulled across the antenna strung between the masts and cut through it. Afterwards, the reconnaissance aircraft would stop the ship, using its on-board weapons dropping bombs before the ships bow if necessary, until the *Hilfkreuzer* arrived on scene.

Above: RC+HR in light grey colors.

Below: This photo of an Ar 199 KK+B? was taken from the cockpit of a chase plane during a training flight near Bug/Rügen in May 1944. Unlike the Ar 196, this aircraft was of all-metal construction, its engine was an Argus As 410c, its wingspan 12.70 meters, length 10.57 meters, height 4.36 meters. Its maximum speed was 260 km/h and could be catapult launched. With instructor and student sitting next to one another, a third crewman sitting behind them could operate the radio equipment.

The sole foreign customer was Bulgaria, which received 12 Ar 196 A-3's in 1943. These were called the "Akula" (shark). The last aircraft of this type, which flew until 1955, today belongs to the collection of the Naval Museum in Varna. An additional two Ar 196's still exist in the U.S. They were taken over at the end of the war as on-board aircraft belonging to the *Prinz Eugen* and served in the testing of the German catapult launching system before being stored for museum purposes, while their mother ship was sunk during testing of the atomic bomb in the Bikini Atoll. The other allied powers of WWII also tested a few of the Arado seaplanes. However, it appears that the three aircraft named above in the U.S., and Bulgaria are the only Arado Ar 196's remaining today.

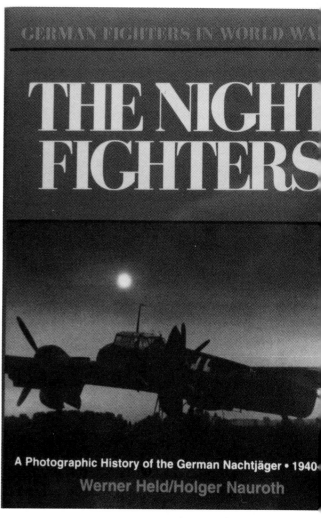